The Coming of

Jetson Thor

*How Nvidia's Next-Gen Robotics Platform is Shaping
the Future of AI and Automation*

Colleen Dutton

Copyright © 2025 by Colleen Dutton

All rights reserved. No part of this book may be reproduced, stored in a retrieval system, or transmitted in any form or by any means—electronic, mechanical, photocopying, recording, or otherwise—without the prior written permission of the author, except for brief quotations in critical reviews or articles.

Disclaimer

The information provided in **The Coming of Jetson Thor: How Nvidia's Next-Gen Robotics Platform is Shaping the Future of AI and Automation** is for educational and informational purposes only. This book is not affiliated with, endorsed by, or sponsored by Nvidia Corporation, Amazon, Tesla, Google, or any other company mentioned within. All trademarks, product names, and company names are the property of their respective owners.

This book does not intend to sell, promote, or advertise any specific product or service. The content reflects independent research and the author's personal analysis of industry trends and technological advancements.

Readers are encouraged to conduct their own research and consult with industry professionals when making decisions based on the information provided in this book. The author and publisher make no representations or warranties regarding the accuracy or completeness of the information contained herein and disclaim any liability for errors, omissions, or outcomes resulting from the application of the concepts discussed.

Table of Contents

Introduction..6

Chapter 1: The Rise of Humanoid Robotics......................14
    A Brief History of Humanoid Robots................................17
    Key Players in the Robotics Industry...............................20
    The Role of AI in Robotics Evolution...............................23

Chapter 2: Nvidia's Foray into Robotics............................27
    Nvidia's Legacy in AI and Computing................................29
    From GPUs to Jetson: A Strategic Shift............................32
    Partnerships and Industry Influence.................................34

Chapter 3: Jetson Thor – A Technical Overview................37
    What is Jetson Thor?..39
    Core Features and Capabilities...42
    How Jetson Thor Stands Apart from Competitors............44

Chapter 4: The Role of Generative AI in Robotics............48
    The Explosion of Generative AI Models............................50
    How Robots Learn and Adapt Through AI.......................53
    Simulation Environments and Robotics Training.............56

Chapter 5: Industry Applications of Jetson Thor...............60
    Manufacturing and Industrial Automation........................63
    Healthcare and Service Robotics......................................66

  Consumer and Personal Robotics.. 69

Chapter 6: Nvidia's OEM Strategy.. 72

  Supplying the Tools, Not Building the Machines................. 75

  Collaboration with Global Robot Makers............................. 78

  The Market for "Hundreds of Thousands" of Robots.......... 80

Chapter 7: Competitive Landscape... 83

  Tesla, Amazon, and Google – The Major Players.................. 86

  How Competitors Are Responding to Nvidia....................... 90

  Will Nvidia Remain the Dominant Force?............................ 93

Chapter 8: The Future of Robotics and AI........................... 96

  What Comes Next for Humanoid Robots?........................ 100

  AI's Role in Shaping Everyday Life...................................... 103

  Predictions for the Next Decade.. 106

Chapter 9: Challenges and Ethical Considerations............ 109

  The Ethics of AI-Powered Robots.. 111

  Addressing Job Displacement and Automation.................. 114

  Ensuring Safety and Accountability..................................... 116

Conclusion... 119

Introduction

The field of robotics has witnessed extraordinary growth over the past few decades, with technological advancements that were once the stuff of science fiction now becoming reality. From industrial automation to the emergence of humanoid robots, the potential applications of robotics in both commercial and consumer markets are vast. At the forefront of this revolution is Nvidia, a company renowned for its innovation in the realm of graphics processing units (GPUs) and artificial intelligence (AI). Through its new offering, **Jetson Thor**, Nvidia is positioning itself as a key player in the robotics ecosystem, providing the hardware that will power the next generation of AI-powered robots.

For years, Nvidia has been synonymous with gaming and graphics, but its focus has gradually shifted toward artificial intelligence and deep learning, which are increasingly being integrated into a variety of industries. The company has become the go-to provider of GPUs for AI research, and its products now underpin many of the most advanced autonomous systems in existence. In robotics, Nvidia has already made significant strides with its Jetson platform, which offers powerful computing solutions for everything from drones to robots. However, with the launch of **Jetson Thor**, the company is taking a more direct step into the world of humanoid robots, where generative AI and high-performance computing are expected to drive the next wave of innovation.

The purpose of this book is to explore how **Jetson Thor** is poised to play a pivotal role in this technological transformation. By understanding the components and capabilities of Jetson Thor, as well as the strategic vision behind its development, readers will gain insights into how this powerful platform is helping to shape the future of robotics. We will also explore the broader implications of Nvidia's push into robotics, particularly as the company continues to supply the underlying technology to hundreds of thousands of robot makers around the world. This shift is not just about hardware—it's about creating the conditions for a revolution in the way robots interact with the world.

The history of robotics is marked by rapid technological advancements, but we are now entering what can only be described as a golden age. Humanoid robots, which were once a distant dream, are now on the cusp of mainstream adoption, largely due to breakthroughs in artificial intelligence and machine learning. These robots are designed to interact with humans in a way that mimics natural behavior, making them valuable in a wide range of industries, from manufacturing to healthcare, logistics, and beyond.

This new era of robotics is not merely about automating tasks traditionally performed by machines; it is about creating robots that can learn, adapt, and even think autonomously. These robots are becoming smarter, more capable, and more human-like than ever before. However, for humanoid robots to achieve their full potential, they require more than just sophisticated algorithms. They need the kind of powerful computational platforms that can handle the demands of AI processing, real-time decision-making, and sensory input—all while being small and efficient enough to be embedded into robots that can operate in real-world environments.

Nvidia's **Jetson Thor** is a game-changer in this context. By leveraging the company's expertise in AI and high-performance computing, **Jetson Thor** offers a compact yet powerful solution that allows humanoid robots to process vast amounts of data and execute complex tasks in real-time. With its ability to train robots using simulated environments, Nvidia is making it possible for these robots to improve and evolve without the need for physical trials. This represents a monumental leap forward in the development of autonomous systems that can operate with minimal human intervention, adapt to new scenarios, and interact more intuitively with the world around them.

As we enter this new era, the boundaries between science fiction and reality continue to blur. Robotics is no longer confined to specialized labs or factory floors. Robots are beginning to become integrated into everyday life, and their capabilities are expanding rapidly. **Jetson Thor** will be a major catalyst in this transformation, enabling robots to evolve from simple automation tools to highly intelligent systems capable of working alongside humans in increasingly complex roles.

The launch of **Jetson Thor** is a significant milestone in the evolution of robotics. But what exactly makes this platform so important, and why should the industry and tech enthusiasts take note? To understand its significance, it is essential to grasp the broader context of the robotics industry and Nvidia's unique position within it.

At the core of the **Jetson Thor** platform is Nvidia's powerful hardware, which is designed to handle the immense computational demands of AI and robotics. With its high-performance GPUs and custom processors, **Jetson Thor** is capable of running complex AI models and algorithms that are essential for tasks such as object recognition, decision-making, and natural language processing. This level of computational power is crucial for humanoid robots, which need to process large amounts of sensory data and interact with their environment in real-time. The platform's ability to simulate environments and train robots virtually is also a game-changer, as it reduces the need for costly and time-consuming physical trials, allowing developers to refine robot behaviors in a digital space before deploying them in the real world.

Another key reason **Jetson Thor** matters is its role in Nvidia's strategic vision for the future of robotics. Rather than competing directly with robot manufacturers, Nvidia aims to become the underlying technology provider for the growing number of robotics companies around the world. By offering a comprehensive platform that includes hardware, software, and AI tools, Nvidia is positioning itself as the backbone of the robotics ecosystem, enabling developers to create robots that can learn, adapt, and scale quickly. The company is providing the infrastructure that makes it easier for robot makers to focus on innovation and user experience, without having to worry about building the computational backbone from scratch.

Moreover, the launch of **Jetson Thor** signals a major shift in the way robots will interact with the world. As AI becomes more advanced and robots become more capable, we are likely to see an explosion of humanoid robots in industries ranging from healthcare and hospitality to logistics and personal assistants. With its scalable and powerful platform, Nvidia is providing the tools that will allow these robots to become smarter, more efficient, and more adaptable. This is the beginning of a new chapter in robotics—one where the robots of tomorrow are not only capable of performing tasks, but also capable of learning, evolving, and interacting with humans in ways we have never seen before.

In short, **Jetson Thor** represents a quantum leap forward in the world of robotics, providing the necessary infrastructure for the next generation of intelligent, autonomous humanoid robots. It is an essential part of

Nvidia's broader strategy to revolutionize AI and robotics, and it will play a key role in shaping the future of automation.

Chapter 1: The Rise of Humanoid Robotics

The rise of humanoid robotics has been one of the most significant technological advancements of the 21st century, driven by a combination of scientific breakthroughs, growing demand for automation, and rapid progress in artificial intelligence (AI) and machine learning (ML). Humanoid robots, designed to resemble human beings in appearance and behavior, are no longer confined to the realm of science fiction. Over the past few decades, these robots have evolved from simple prototypes into advanced machines capable of performing complex tasks, interacting with humans, and adapting to dynamic environments.

At the core of the rise of humanoid robots is the rapid advancement in AI and machine learning, which allows these robots to interpret sensory input, make decisions, and perform actions with a level of autonomy that was once thought to be unattainable. Additionally, the integration of sophisticated sensors, advanced actuators, and powerful computational platforms like Nvidia's Jetson Thor is enabling humanoid robots to achieve a level of dexterity, mobility, and cognitive function that can mimic human behavior.

Humanoid robots are being developed for a wide range of applications, from healthcare and manufacturing to customer service and entertainment. In healthcare, robots like Japan's Robear are designed to assist elderly patients with mobility and caregiving tasks, responding to their needs in real-time. In manufacturing, humanoid robots are beginning to play roles in assembly lines, performing intricate tasks alongside human workers. Tesla's Optimus robot, currently under development, is expected to assist with household chores, demonstrate dexterity, and even navigate complex environments.

The potential of humanoid robots lies in their ability to interact with the physical world in ways that traditional robots, such as industrial robots, cannot. These robots are designed with anthropomorphic characteristics—arms, legs, and faces—to perform tasks that require dexterity, mobility, and social interaction. With humanoid robots capable of navigating diverse environments and interacting with humans, industries ranging from healthcare to entertainment will experience profound changes in their operations.

Moreover, humanoid robots are becoming increasingly versatile thanks to their integration with advanced AI algorithms. These robots can learn from their interactions, improving their abilities over time. The technology behind humanoid robotics has made it clear that the future of automation is not just about machines doing repetitive tasks but rather about machines that can understand, learn, and adapt in human-centric environments.

While the pace of humanoid robotics development is accelerating, it is important to note that challenges remain. Issues such as human-robot interaction, safety, and the ethical implications of humanoid robots in the workplace continue to be important areas of study and regulation. Nevertheless, the rise of humanoid robots promises to be transformative, and as AI and robotics technologies continue to evolve, we can expect these machines to increasingly become part of our daily lives.

A Brief History of Humanoid Robots

The concept of humanoid robots has fascinated humanity for centuries, from ancient myths of artificial beings to the development of mechanical automatons in the 20th century. The history of humanoid robotics is a testament to human creativity, engineering ingenuity, and the enduring desire to create machines that mimic human capabilities.

The earliest examples of humanoid robots can be traced back to the 15th century, with inventors like Leonardo da Vinci conceptualizing mechanical human-like machines. Da Vinci's famous sketch of a mechanical knight in 1495 is considered one of the earliest ideas for a humanoid robot. While it was never built, the concept was a precursor to later developments in robotics.

The 18th century saw the creation of more advanced automatons, often referred to as "mechanical men." These machines, such as those built by Swiss watchmaker Jaquet-Droz, were capable of performing simple tasks, like writing or drawing, and represented an early attempt to replicate human actions mechanically. While these machines were not fully autonomous, they laid the groundwork for the more advanced humanoid robots of the 20th century.

The 20th century brought significant advancements in both the technology and concept of humanoid robots. One of the most famous examples from this period was the creation of "The Unimate" in 1961 by George Devol and Joseph Engelberger. While The Unimate was not humanoid in the traditional sense, it was the first industrial robot to be installed on an assembly line. This marked the beginning of modern robotics, but humanoid robots were still a distant dream.

The first true humanoid robot, as we understand the term today, emerged in the late 20th century. In 1997, Honda introduced its first iteration of ASIMO (Advanced Step in Innovative Mobility), a bipedal humanoid robot capable of walking, climbing stairs, and interacting with people. ASIMO became one of the most iconic humanoid robots, showcasing the potential for robots to perform human-like tasks.

The 21st century witnessed rapid advancements in humanoid robotics. In 2016, Hanson Robotics unveiled Sophia, a humanoid robot designed to simulate human facial expressions and engage in conversations. Sophia's ability to interact with people, answer questions, and express emotions captured the world's attention and highlighted the growing capabilities of humanoid robots in human-robot interactions.

Since then, numerous companies and research institutions have developed humanoid robots with enhanced mobility, autonomy, and social intelligence. Tesla's Optimus robot, which aims to perform tasks in both the home and workplace, and SoftBank's Pepper robot, designed for customer service, are some of the key players in this field today.

As the technology behind humanoid robots continues to advance, the dream of creating robots that can seamlessly integrate into human environments becomes increasingly achievable. However, as this field matures, it raises important ethical and social questions about the role of robots in society and their impact on the workforce.

Key Players in the Robotics Industry

The robotics industry is vast and diverse, with key players spanning several sectors, including manufacturing, healthcare, autonomous vehicles, and consumer products. The rise of humanoid robots has drawn attention to a select few companies and research organizations that are at the forefront of developing these advanced technologies. While traditional robotics companies like KUKA and ABB continue to dominate industrial robotics, new companies focused on humanoid robots and AI-driven automation have emerged, shaping the future of the industry.

1. **Nvidia**: Nvidia has established itself as a leader in the AI and robotics space, thanks to its powerful GPUs and deep learning technologies. With the introduction of the Jetson platform, Nvidia is enabling developers to create robots with advanced AI capabilities. The company's work with humanoid robots, such as its involvement with Tesla's Optimus robot, shows its significant impact on the field.

2. **Boston Dynamics**: Known for creating some of the most advanced robots in the world, Boston Dynamics is a leader in the development of mobile robots. While its robots, such as Spot and Atlas, are not strictly humanoid, they are highly mobile, capable of performing complex tasks, and represent the cutting-edge of robotics.

3. **Honda**: Honda's ASIMO robot, introduced in 2000, is one of the most iconic humanoid robots ever created. ASIMO is capable of walking, running, climbing stairs, and interacting with people. Honda's continued work in this field has made it a key player in humanoid robotics, although the company has shifted its focus to other areas, such as mobility solutions.

4. **SoftBank Robotics**: Known for its Pepper robot, which is designed to interact with humans in social settings, SoftBank Robotics has been a pioneer in creating humanoid robots for commercial use. Pepper is used in retail environments, hotels, and healthcare settings, demonstrating the potential for humanoid robots to perform customer service tasks.

5. **Tesla**: Tesla's foray into humanoid robotics with the development of Optimus represents a bold move into a new market. While still in its early stages, the company's focus on creating a

humanoid robot for household and industrial tasks shows the potential for robotics to revolutionize multiple industries.

6. **Agility Robotics**: Agility Robotics has developed Cassie, a bipedal robot capable of running and walking. The company's focus on creating robots that are not only humanoid but also highly functional in a variety of environments positions it as a key player in the field of advanced robotics.

These companies, along with several universities and research institutions, are pushing the boundaries of what humanoid robots can do, each with a unique focus on advancing mobility, AI, and human-robot interaction.

The Role of AI in Robotics Evolution

The role of artificial intelligence in the evolution of robotics is central to the current and future capabilities of robots. AI empowers robots to learn from their environments, adapt to new tasks, and perform complex actions autonomously. Over the past few decades, the integration of AI into robotics has revolutionized the field, making robots more intelligent, versatile, and capable of interacting with humans in increasingly sophisticated ways.

AI technologies such as machine learning, computer vision, natural language processing, and reinforcement learning are enabling robots to perform tasks that were once thought to be exclusive to humans. Machine learning allows robots to improve their performance through experience, while computer vision helps robots interpret and navigate the physical world by recognizing objects, faces, and gestures. Natural language processing enables robots to understand and respond to human language, making them more effective in customer service and healthcare applications.

Reinforcement learning, a subset of machine learning, has proven especially important for robotics. This technology allows robots to learn through trial and error, optimizing their actions based on feedback from the environment. For example, robots like Atlas from Boston Dynamics use reinforcement learning to improve their movement and agility, while humanoid robots can learn to perform tasks such as assembling products or interacting with customers.

AI has also played a crucial role in enabling robots to perform in dynamic and unpredictable environments. This is particularly evident in autonomous robots, such as self-driving cars and delivery drones, which rely heavily on AI to make real-time decisions based on sensor input and environmental data. In humanoid robotics, AI is enabling robots to better understand and react to human emotions, facial expressions, and social cues, making them more capable of interacting with people in a meaningful way.

As AI continues to evolve, its integration into robotics will open up new possibilities for automation, efficiency, and human-robot collaboration. The future of humanoid robots will undoubtedly rely on AI advancements, as robots will be able to learn, adapt, and perform tasks with greater precision and autonomy than ever before.

Chapter 2: Nvidia's Foray into Robotics

Nvidia's journey into robotics is both a natural extension of its core capabilities in AI and computing and a strategic pivot towards the next frontier of technology. While the company is primarily known for its Graphics Processing Units (GPUs), which have been instrumental in revolutionizing the gaming industry, Nvidia has steadily expanded its reach into fields like data centers, artificial intelligence (AI), and, more recently, robotics. This move into robotics, epitomized by the launch of its Jetson platform and now the upcoming Jetson Thor, represents a convergence of AI, deep learning, and robotics—technologies that are becoming increasingly intertwined in modern automation and intelligent systems.

Robotics is a sector with enormous growth potential, and Nvidia's strategic investment in this area reflects both market demand and a broader vision of transforming industries through automation. The Jetson platform was launched to serve the growing need for compact, low-power, yet high-performance computing solutions in robotics. Nvidia's primary aim is not to compete with robotics manufacturers directly but to provide them with the tools and technologies necessary to build advanced robots capable of navigating complex environments, making autonomous decisions, and learning from their surroundings.

The company has recognized that the future of robotics lies not just in hardware, but in the software and intelligence that power it. With advancements in AI, particularly in machine learning and computer vision, robots can now be trained to interact with their environments in ways that were once thought impossible. Nvidia's AI hardware, especially its GPUs, has enabled deep neural networks and other advanced algorithms to function at unprecedented scales. By leveraging its expertise in AI and computing, Nvidia is positioning itself as the backbone provider of intelligent systems for robotics, providing everything from hardware to software stacks that facilitate deep learning and reinforcement learning in real-time robotic applications.

This push into robotics is also strategically timed. As industries like manufacturing, healthcare, and logistics begin adopting robotic systems at scale, Nvidia stands poised to deliver the necessary computational power to make these machines smarter and more autonomous. Whether it's an autonomous delivery robot navigating crowded streets, a warehouse robot optimizing its movement to improve productivity, or a healthcare robot assisting in surgeries, Nvidia's technology underpins many of these breakthroughs. This ability to deliver the necessary computing power has positioned Nvidia as a key player in the evolving robotics ecosystem.

Nvidia's Legacy in AI and Computing

Nvidia has long been a leader in the world of computing, particularly in the realm of graphics processing. The company's legacy in AI and computing is inextricably tied to its innovations in GPUs, which were initially designed for rendering high-quality graphics in video games. However, as the need for greater computational power grew—especially in the fields of deep learning and AI—Nvidia recognized the potential of its GPUs to accelerate AI workloads far beyond the gaming industry.

The introduction of CUDA (Compute Unified Device Architecture) in 2006 was a pivotal moment in Nvidia's history. CUDA is a parallel computing platform and application programming interface (API) model that enables software developers to harness the massive parallel computing power of GPUs for non-graphical workloads. This move opened up new possibilities for AI, scientific research, and high-performance computing (HPC), enabling industries to achieve computational speeds that were previously unthinkable.

With the rise of machine learning and deep learning, Nvidia's GPUs quickly became the go-to solution for researchers, data scientists, and AI practitioners. The company's hardware accelerated the training of deep neural networks, which are the backbone of many AI systems today. As deep learning algorithms began to solve complex problems in image recognition, natural language processing, and even game-playing, Nvidia was right there, providing the computational muscle that made it all possible.

Nvidia's dominance in AI was further solidified with the release of the Tesla and Quadro series of GPUs, designed specifically for professional workstations and data centers. The company's focus on AI and computing continued with the launch of the DGX platform—a complete system built for AI research and development, incorporating Nvidia's GPUs and software. This shift from gaming graphics to AI and computing at scale cemented Nvidia's reputation as a leader in the AI hardware space.

As AI exploded into mainstream industries—transforming sectors like healthcare, finance, and logistics—Nvidia continued to innovate. Its GPUs became integral to the development of autonomous systems, robotics, and AI-powered applications across the board. This evolution from graphics to general-purpose AI computing has allowed Nvidia to establish itself as not just a hardware provider, but a foundational technology company for the entire AI ecosystem.

From GPUs to Jetson: A Strategic Shift

Nvidia's strategic shift from a purely gaming-centric company to a leader in AI and robotics has been nothing short of transformative. The introduction of the Jetson platform marked a turning point for the company. While Nvidia's GPUs revolutionized AI and deep learning in large data centers, the Jetson platform was designed to bring the power of GPUs to edge devices—compact systems that could be deployed in real-world environments such as robots, drones, and other intelligent devices.

The Jetson platform is a direct response to the need for more efficient, compact, and low-power solutions for embedded systems. Unlike traditional data centers, these edge devices require hardware that can process AI workloads locally without relying on constant internet connectivity. Nvidia's answer to this challenge was the Jetson line of products, which integrate high-performance GPUs with CPUs, memory, and other necessary components in a single, small form factor. These systems are designed to be energy-efficient, enabling AI-powered devices to perform complex computations on the edge, while also being flexible enough to support a variety of robotics, AI, and automation use cases.

Jetson's impact on robotics cannot be overstated. The platform enables developers to create autonomous systems capable of performing sophisticated tasks in environments where traditional computing power is unavailable or impractical. Whether it's enabling autonomous drones to navigate complex environments, robots performing repetitive tasks in warehouses, or industrial automation, Jetson has become the engine behind a new generation of smart devices.

The strategic shift from GPUs to the Jetson platform represents Nvidia's broader vision of being at the heart of the next wave of technological innovation. By moving into robotics, Nvidia has broadened its reach beyond the data center, positioning itself as a leader in edge computing and AI-powered automation. This shift also reflects the growing demand for more intelligent, autonomous systems capable of solving real-world problems in industries such as manufacturing, logistics, healthcare, and beyond.

Partnerships and Industry Influence

Nvidia's influence in the tech industry extends far beyond its own products. The company has built an extensive network of partnerships with key players across various industries, cementing its position as a leader in AI and robotics. These partnerships have allowed Nvidia to expand the reach of its technology, integrate its products into a diverse array of applications, and shape the future of computing, robotics, and automation.

One of Nvidia's most important partnerships is with Tesla, which uses Nvidia's GPUs and AI hardware to power its Autopilot system and the development of its humanoid robot, Optimus. This collaboration is a prime example of how Nvidia's technology underpins cutting-edge innovations in autonomous systems and robotics. Tesla's reliance on Nvidia's AI hardware highlights the critical role the company plays in enabling the development of autonomous vehicles and robots.

Beyond Tesla, Nvidia has cultivated partnerships with major players in industries such as healthcare, retail, and logistics. Its work with companies like Amazon, Google, and Microsoft has helped drive the adoption of AI-powered technologies across the enterprise. Nvidia's GPUs are integral to the deep learning workloads that power everything from Amazon's AWS services to Google's AI research and development.

The company's partnerships with academic institutions and research organizations have also been instrumental in advancing the state of AI and robotics. Nvidia collaborates with universities around the world, providing researchers with the computational tools needed to push the boundaries of AI, deep learning, and robotics. These collaborations have not only helped Nvidia stay at the forefront of technological development but have also ensured that its products remain central to the evolution of AI.

Nvidia's foray into robotics, supported by its legacy in AI and computing, its strategic shift to Jetson, and its deep industry partnerships, positions the company as a central player in the future of intelligent systems and automation. Through its innovative technologies and collaborations, Nvidia is shaping the landscape of robotics and AI, making them more accessible, intelligent, and capable of solving real-world challenges.

Chapter 3: Jetson Thor – A Technical Overview

The Jetson Thor is Nvidia's next-generation robotics platform, representing a significant leap in the company's strategic push to dominate the AI-powered robotics space. With a focus on providing robust computational power for humanoid robots, Jetson Thor is built on Nvidia's extensive experience in AI, deep learning, and edge computing. This platform is positioned as a crucial enabler of intelligent robotics applications in industries ranging from manufacturing and logistics to healthcare and personal robotics.

At its core, Jetson Thor integrates Nvidia's most advanced AI models, including generative AI capabilities, into compact, energy-efficient hardware designed to support the next generation of humanoid robots. These robots are tasked with performing complex tasks, such as interacting with humans, navigating dynamic environments, and learning from real-time data.

Jetson Thor is optimized to leverage Nvidia's cutting-edge technologies, such as CUDA (Compute Unified Device Architecture), Tensor Cores, and deep learning accelerators, providing developers with a platform that can handle the computational requirements of autonomous robots. These robots are typically required to process vast amounts of sensor data in real-time, make decisions, and adapt to dynamic environments, all of which necessitate powerful processing capabilities.

The system-on-module (SoM) design of Jetson Thor allows for high-performance processing without compromising on the size and weight constraints typical of humanoid robotics. As a result, Jetson Thor can be embedded into compact robotic platforms, providing them with the computational horsepower required to function efficiently in real-world applications.

Jetson Thor operates within Nvidia's broader AI and robotics ecosystem, integrating seamlessly with Nvidia's AI models, including GPT-based architectures and other generative models. These models are key to enabling robots to understand and interact with their surroundings in a more human-like manner. The ability to leverage simulated environments for robot training is another significant advantage that Jetson Thor provides, making it easier for developers to train robots in virtual environments before deploying them in the real world.

What is Jetson Thor?

Jetson Thor is a cutting-edge AI computing platform developed by Nvidia, specifically designed to power humanoid robots with advanced capabilities. The platform builds on Nvidia's deep expertise in AI, edge computing, and high-performance computing hardware to create a solution that enables humanoid robots to perform a wide range of tasks autonomously and intelligently.

The platform itself is a compact, modular system built around Nvidia's powerful GPUs, which have long been the backbone of AI processing. Jetson Thor integrates these GPUs with specialized processors to enable real-time processing of large amounts of sensor data—critical for tasks like navigation, object recognition, and human interaction. This integration allows humanoid robots to make quick decisions, learn from their surroundings, and interact with people in increasingly sophisticated ways.

Jetson Thor's purpose is to provide developers with the tools needed to build highly functional, autonomous robots that can operate in dynamic environments. It leverages Nvidia's hardware accelerators, such as Tensor Cores and CUDA, to deliver the computational power required for real-time AI tasks. These tasks include deep learning, computer vision, and natural language processing, which are fundamental to enabling humanoid robots to understand their surroundings and perform tasks intelligently.

Additionally, Jetson Thor's hardware is optimized for edge computing, meaning it can process data locally, on the robot itself, without needing to rely on a constant connection to a cloud server. This not only reduces latency but also makes robots more adaptable to environments where reliable internet connectivity may be unavailable or inconsistent.

The system's design also emphasizes low power consumption, which is essential for humanoid robots that need to operate for extended periods without needing frequent recharging. This makes Jetson Thor particularly suitable for industries like logistics, healthcare, and public services, where humanoid robots are expected to perform tasks in the field over long durations.

Jetson Thor is part of Nvidia's broader strategy to provide AI and computing infrastructure that powers the next wave of robotics innovation. It is designed to complement Nvidia's other platforms, such as the Jetson Nano and Jetson Xavier, which target other robotics segments like drones and industrial robots. However, Jetson Thor stands out due to its specific focus on humanoid robots, positioning it as the premier platform for next-gen robotic applications.

Core Features and Capabilities

Jetson Thor offers a range of features and capabilities that make it an exceptional platform for building intelligent humanoid robots. Among its most notable attributes are its processing power, scalability, power efficiency, and developer-friendly environment.

Processing Power and AI Capabilities

At the heart of Jetson Thor's computational abilities is Nvidia's GPU architecture, which is designed to excel in AI tasks. The platform integrates high-performance GPUs capable of handling massive data inputs from sensors like cameras, LIDAR, and depth sensors. This allows the robot to perceive its environment, understand object dynamics, and make real-time decisions.

Jetson Thor's use of Tensor Cores allows it to accelerate deep learning and neural network processing, enabling robots to perform advanced tasks like facial recognition, natural language processing, and predictive maintenance. It also supports generative AI models, which can allow robots to autonomously create new strategies or solutions based on the data they process. This makes Jetson Thor ideal for humanoid robots that need to operate autonomously and adapt to ever-changing environments.

Edge Computing and Real-Time Processing

Jetson Thor leverages edge computing, which enables processing directly on the robot without relying on cloud-based computing. This is particularly important for humanoid robots, as it reduces latency and ensures that robots can respond quickly to their environment. For example, a robot performing delivery tasks in a warehouse can navigate obstacles, avoid collisions, and plan routes without waiting for cloud updates.

By performing data processing locally, Jetson Thor ensures that robots can operate reliably even in environments with limited internet connectivity. This capability is vital in industries like healthcare, where robots need to interact with patients in real-time without delays.

Scalability and Flexibility

Jetson Thor's modular design allows for easy scalability, making it suitable for a range of robotic platforms, from lightweight consumer robots to heavy-duty industrial machines. This scalability ensures that Jetson Thor can support everything from entry-level robots to more complex humanoid systems.

Additionally, Jetson Thor is flexible enough to accommodate various AI frameworks, including TensorFlow, PyTorch, and Nvidia's own DeepStream SDK, providing developers with a diverse toolkit to build custom solutions. Its compatibility with Nvidia's simulation tools, such as Isaac Sim, enables developers to train robots in virtual environments, reducing the cost and risk associated with physical trial and error.

How Jetson Thor Stands Apart from Competitors

While Jetson Thor is not the only AI platform for robotics, it distinguishes itself through several key features, from computational power to developer tools. Here's how it stands apart from the competition.

Unmatched AI and GPU Processing Power

One of the primary differentiators for Jetson Thor is Nvidia's unrivaled expertise in AI and GPU technology. Unlike many competitors in the robotics field, Nvidia has years of experience developing GPUs and AI accelerators that are specifically designed to handle the high-throughput, parallel computing required for real-time AI tasks. This allows Jetson Thor to deliver superior processing power for tasks like object detection, navigation, and speech recognition.

Ecosystem and Developer Support

Nvidia's broader ecosystem sets Jetson Thor apart. The platform benefits from Nvidia's robust software libraries, AI models, and simulation tools that are optimized for robotics. The availability of pre-trained models and tools like Isaac SDK and Isaac Sim gives developers a head start, making it easier to build sophisticated robots without starting from scratch.

In comparison, competitors like Intel's Movidius and Qualcomm's robotics platforms offer powerful computing capabilities, but they don't provide the same level of AI optimization or extensive developer ecosystem. Nvidia's deep learning libraries and established presence in the AI space give Jetson Thor a significant edge.

Focus on Humanoid Robotics

While Nvidia's Jetson platform can be used in a variety of robotic applications, Jetson Thor is uniquely tailored to meet the demands of humanoid robotics. Its compact form factor, specialized processing capabilities, and low power consumption make it ideal for humanoid robots designed to operate in human environments. This focused approach gives Jetson Thor a distinct advantage over general-purpose robotics platforms, which might not be optimized for the unique challenges posed by humanoid robots.

These features and distinctions ensure that Jetson Thor is not just another robotics platform, but a pivotal technology that will define the next generation of humanoid robots, enabling them to perform in a wide range of real-world scenarios.

Chapter 4: The Role of Generative AI in Robotics

Generative AI, a subset of artificial intelligence, has been revolutionizing the field of robotics in recent years. While traditional AI focuses on specific tasks, such as recognition, classification, or decision-making, generative AI goes a step further by generating new content, strategies, or solutions from learned data. In robotics, this capability is proving to be a game-changer in both development and functionality.

At the heart of generative AI's role in robotics is its ability to create novel solutions to problems that robots face in dynamic environments. Humanoid robots, industrial robots, and service robots all interact with the world in highly complex ways, requiring real-time adaptation to changing conditions. Generative AI enables robots to not only perform tasks but also to enhance their problem-solving abilities through learned experiences. For example, instead of programming a robot with a fixed set of instructions for tasks like picking up objects, generative AI allows the robot to generate a new set of actions in response to different objects or environments.

Moreover, generative AI is crucial in making robots more autonomous. By generating novel strategies for navigation or manipulation based on experience, these robots can evolve over time, improving their own performance without explicit reprogramming. This is essential for applications in fields like manufacturing, healthcare, and autonomous vehicles, where robots must constantly adapt to new scenarios that may not have been anticipated in their original programming.

Another vital role of generative AI in robotics is its integration with reinforcement learning (RL). In RL, robots learn by trial and error, receiving feedback on actions taken. Generative AI models can enhance this process by generating diverse scenarios and tasks that challenge robots in ways a static dataset cannot. By creating simulated environments with generative AI, robots can train on an infinite set of possible conditions, ultimately leading to more robust and flexible autonomous systems.

The ability of generative AI to create realistic, plausible outcomes from partial data is also a breakthrough for training robots in complex environments. Traditionally, robotics required physically simulating environments, which could be costly and time-consuming. Generative AI reduces the need for extensive physical setups, allowing virtual models and simulations to be generated quickly and with remarkable accuracy. This has drastically reduced the time to market for new robotics applications, enabling faster iterations in product design, testing, and deployment.

The Explosion of Generative AI Models

The explosion of generative AI models marks a defining moment in the AI and robotics industries. Historically, AI was largely focused on solving specific problems through supervised learning, such as classification tasks where algorithms are trained on labeled datasets to recognize patterns or make decisions. However, the development of generative AI models has unlocked a new dimension of machine learning, where AI systems are not just passive learners but active creators.

Generative AI models are powered by deep learning techniques, particularly neural networks known as generative adversarial networks (GANs) and variational autoencoders (VAEs). These models are trained on vast datasets, but instead of simply categorizing data, they generate entirely new data points that resemble the training set. For instance, a generative model trained on a large dataset of images could generate entirely new images that look strikingly similar to the original dataset, even though they've never been seen before.

This capability has created ripples across multiple industries, including robotics. By being able to generate new content, strategies, or environments, generative AI models have vastly improved the performance and adaptability of robots. In the context of robotics, generative models allow machines to simulate and experiment with solutions to problems in environments that might otherwise be difficult or expensive to create. This revolution is evident in sectors ranging from automated content creation to virtual testing of robotic systems.

Another key development in generative AI has been the rise of transformer-based models, which have achieved breakthroughs in language processing, image generation, and even multimodal applications. The recent explosion of large language models (LLMs), such as GPT-3, has demonstrated the power of generative AI to understand and produce human-like language, an attribute that can extend to robots interacting with people. With advancements like these, generative AI is being harnessed for robots to improve both their communication capabilities and decision-making processes. For instance, humanoid robots can now generate more realistic responses to questions and interact with humans in a more natural way, thanks to generative AI models trained on extensive conversational data.

Moreover, the scalability of generative AI models has made them more accessible to companies of all sizes. Previously, building and training advanced AI models was the domain of large tech giants like Google or Microsoft. Today, the availability of powerful pre-trained models, combined with the rise of cloud computing platforms, allows smaller robotics companies to adopt and customize these tools for their own needs. This democratization of AI technology has accelerated innovation in robotics, paving the way for more diverse applications across industries such as healthcare, logistics, and entertainment.

How Robots Learn and Adapt Through AI

Robots' ability to learn and adapt through AI has been one of the most critical factors in their growing autonomy and success. Unlike traditional machines, which operate based on fixed instructions and cannot adjust to unforeseen circumstances, AI-powered robots can adapt to their environments and improve over time.

This adaptability is primarily due to the application of machine learning, particularly deep learning and reinforcement learning (RL), which allows robots to continuously improve their performance based on feedback.

Deep learning, a subset of machine learning, enables robots to process vast amounts of unstructured data, such as images, sounds, and even tactile sensations, to make intelligent decisions. By using neural networks with multiple layers, robots can learn complex patterns in data that would be difficult or impossible for humans to manually program. This process is known as supervised learning, where robots are trained on labeled datasets to learn the correlation between inputs and desired outputs.

Reinforcement learning, on the other hand, allows robots to learn through trial and error. In RL, robots interact with their environment and receive feedback in the form of rewards or penalties based on their actions. This feedback loop helps the robot adjust its behavior to maximize the reward, allowing it to perform tasks more effectively. Over time, the robot learns to make better decisions, refine its strategies, and become more efficient at completing tasks. This kind of learning is crucial for robots that must operate in unpredictable and dynamic environments, such as autonomous vehicles or service robots that interact with humans.

Furthermore, robots can employ a technique called transfer learning, which allows them to apply knowledge gained from one task to a new, similar task. For example, a robot trained to sort packages based on size can transfer its learning to sort packages based on weight. This helps robots learn faster and with less data, making them more adaptable in new situations.

By continually learning from new data and experiences, robots can adapt to environmental changes, improve their performance, and even develop new behaviors that were not explicitly programmed. This is essential for industries where robots must handle a variety of tasks and deal with unexpected challenges. Autonomous delivery robots, for instance, can learn to navigate streets, avoid obstacles, and adapt to new delivery routes without the need for manual intervention.

Simulation Environments and Robotics Training

Simulation environments play a pivotal role in modern robotics training, enabling robots to learn and practice tasks in virtual spaces before engaging with the real world. This approach allows engineers to test robots in a controlled setting, eliminating the risks and costs associated with physical trials. The use of simulation has become particularly important in the development of complex AI-powered robots, which require massive amounts of training data and diverse scenarios to function effectively.

One of the key advantages of using simulation for robotics training is the ability to generate a wide range of scenarios that a robot might encounter in real-world environments. For example, autonomous vehicles can be trained to navigate through diverse traffic conditions, weather patterns, and road obstacles without the need for real-world testing, which can be dangerous and expensive. Similarly, humanoid robots can simulate human interactions, facial expressions, and movements, allowing them to learn social cues and behaviors before being deployed in public settings.

Simulation environments are particularly valuable in reinforcement learning, where robots need to experiment and receive feedback on their actions. These environments provide a safe space for robots to practice without causing harm to themselves or others. By running multiple simulations, robots can experience a variety of scenarios, making them more adaptable when exposed to new situations. Furthermore, the use of generative AI models in simulation environments allows for the creation of realistic, dynamic worlds that better reflect real-life challenges.

The development of highly accurate physics engines and realistic virtual environments has enhanced the realism of simulations. These tools allow robots to interact with their environment in a lifelike manner, learning to handle complex tasks such as grasping objects, walking, or navigating through dynamic environments. For instance, simulation platforms like Gazebo, Webots, and NVIDIA's Isaac Sim provide realistic environments for robotic systems to test algorithms and improve their skills.

Simulation also accelerates the training process by allowing for rapid iteration. In traditional robotic training, real-world testing would be slow and costly, especially when faced with the need for large datasets or repeated trials. Simulated environments eliminate this barrier, providing near-instant feedback and reducing the time required for training. This enables faster development cycles, allowing robots to be deployed more quickly and efficiently.

Simulation environments are integral to modern robotics, providing a safe, cost-effective, and scalable way to train robots. Through simulated experiences, robots can improve their skills, adapt to changing environments, and learn complex tasks—ushering in a new era of more capable, intelligent machines.

Chapter 5: Industry Applications of Jetson Thor

Nvidia's Jetson Thor, with its powerful computing platform tailored for robotics, is positioned to revolutionize several key industries by providing sophisticated AI capabilities, enabling robots to learn, adapt, and make decisions in real-time. The introduction of this technology marks a pivotal moment in the development of robotics across sectors such as manufacturing, healthcare, and consumer applications. By equipping robots with the ability to operate autonomously in complex, dynamic environments, Jetson Thor is facilitating the next stage of automation and transforming industries in unprecedented ways.

At the core of Jetson Thor's impact is its ability to support highly advanced AI-driven functionalities, such as generative models, natural language processing, and object recognition, through its specialized hardware and software integrations. This empowers robots to carry out tasks with a degree of intelligence that was previously unimaginable, enabling companies to improve efficiency, reduce costs, and innovate in ways that redefine what's possible.

In manufacturing and industrial automation, Jetson Thor is expected to play a crucial role in streamlining operations. By integrating AI into assembly lines, Jetson-powered robots can not only perform repetitive tasks like assembling components or packaging but also adapt to new tasks with minimal human intervention. These robots are capable of learning and improving their performance through simulations and generative AI, making them invaluable assets to manufacturing facilities. The technology's compact nature makes it ideal for deployment in small, agile robots capable of navigating tight spaces and performing high-precision tasks with ease.

In healthcare, Jetson Thor's capabilities could radically alter the way medical robots assist in surgeries, patient care, and diagnosis. With the ability to process vast amounts of real-time data, Jetson-powered robots can offer surgical assistance with pinpoint accuracy, enhance telemedicine, and even perform rehabilitation exercises for patients with mobility impairments. The precision and intelligence offered by Jetson Thor could pave the way for more efficient, personalized healthcare solutions.

For consumer and personal use, the applications are equally diverse. From home assistants to service robots, Jetson Thor-powered devices can be designed to seamlessly integrate into daily life, offering improved services, greater convenience, and smarter solutions for households. Robots could perform household chores, help elderly individuals with mobility issues, or provide emergency assistance in critical situations, all while learning and improving their interactions with users through AI-powered feedback loops.

As the applications of Jetson Thor expand, it is clear that Nvidia's platform is not just about creating smarter robots but enabling a fundamental shift in how industries interact with automation. With its robust AI capabilities, Jetson Thor is set to become a critical enabler in various sectors, pushing the boundaries of what is possible in robotics.

Manufacturing and Industrial Automation

In manufacturing and industrial automation, the advent of advanced robotics powered by Nvidia's Jetson Thor is reshaping the landscape of production. By integrating cutting-edge AI with robust computing power, Jetson Thor is enabling a level of automation and intelligence previously reserved for high-cost, large-scale operations. The technology's compactness, paired with its real-time data processing abilities, makes it ideal for a wide range of applications, from assembly lines to logistics management.

One of the primary benefits of Jetson Thor in manufacturing is its ability to streamline production processes. Automated robots powered by Jetson Thor can take on repetitive tasks such as material handling, assembly, and packaging. These robots can operate autonomously with minimal supervision, making decisions in real time based on sensory inputs. For example, a Jetson-powered robot on an assembly line can identify faults in components and correct them without the need for human intervention, ensuring high-quality products and reducing downtime caused by manual inspections.

Moreover, Jetson Thor's real-time processing capabilities open up new possibilities for predictive maintenance. In traditional manufacturing environments, equipment failures can cause costly delays. However, with the power of AI and machine learning, Jetson Thor can analyze data from sensors embedded in machinery to predict potential failures before they occur. By flagging issues proactively, robots equipped with Jetson Thor can allow manufacturers to carry out preventative maintenance, avoiding costly shutdowns and optimizing machine performance.

The technology's ability to process large amounts of data from multiple sources simultaneously also enhances the flexibility of manufacturing lines. In a factory where production demands are constantly changing, Jetson-powered robots can quickly adjust to new tasks. For example, if a factory switches from producing one product type to another, the robots can adapt without needing to be reprogrammed or recalibrated, making production lines more agile and responsive to market demands.

Furthermore, Jetson Thor's small form factor makes it ideal for deployment in confined spaces, where traditional automation systems may struggle to fit. These robots can move freely, navigate complex environments, and even interact with human workers safely, adding an additional layer of efficiency and flexibility to industrial operations.

The implementation of Jetson Thor in industrial automation is not just about improving existing processes but enabling entirely new models of manufacturing. From autonomous factories to AI-driven supply chains, the capabilities of Jetson Thor are pushing the boundaries of what's possible, ushering in a new era of smart, efficient, and scalable production.

Healthcare and Service Robotics

Healthcare is one of the most promising areas for the application of Jetson Thor-powered robotics. With its ability to process vast amounts of data in real time, coupled with AI capabilities such as computer vision, natural language processing, and decision-making, Jetson Thor is poised to significantly enhance both patient care and operational efficiency within the healthcare sector.

Surgical robots, for example, are already transforming the field of surgery by allowing for more precise operations with minimal human intervention. Powered by Jetson Thor, surgical robots could go even further by incorporating AI that helps identify potential complications, suggest optimal procedures, and learn from past operations to refine techniques over time. The combination of AI and real-time data processing allows these robots to perform complex surgeries with greater accuracy, reducing the risk of human error and improving patient outcomes.

Additionally, Jetson Thor could play a vital role in telemedicine. With its AI-powered algorithms, robots equipped with Jetson Thor could provide remote consultations, diagnosis, and even assist in physical examinations. For patients in rural or underserved areas, this could mean faster access to medical expertise and more timely interventions. The real-time processing power of Jetson Thor ensures that these remote interactions can be as accurate and reliable as in-person consultations.

In the area of rehabilitation, robots powered by Jetson Thor could assist patients in recovering from injuries or surgeries. AI-driven rehabilitation robots can tailor physical therapy exercises to individual patients, adjusting in real time to their progress and offering real-time feedback. This personalized approach can lead to faster recovery times and better long-term outcomes for patients.

Jetson Thor's capabilities extend to the service robotics industry as well. Robots in hospitals, nursing homes, and clinics can assist with a variety of tasks, from transporting supplies to delivering medications and meals. These robots can navigate through crowded hospital corridors autonomously, freeing up medical staff to focus on more critical tasks. The ability of Jetson Thor to process sensory data quickly allows these robots to interact safely and effectively with patients and medical personnel.

As the healthcare industry continues to embrace automation, Jetson Thor is uniquely positioned to drive innovation, offering more intelligent, adaptive, and precise robots that could improve patient outcomes, streamline healthcare delivery, and create a more efficient medical environment.

Consumer and Personal Robotics

Consumer and personal robotics is an area where the integration of AI and advanced computing, such as that offered by Nvidia's Jetson Thor, is already making waves. Jetson Thor's advanced capabilities open up new possibilities for robots to become integral parts of everyday life, providing enhanced convenience, assistance, and personalized experiences for consumers.

For home automation, Jetson Thor could power robots that handle a variety of household chores, such as cleaning, cooking, and maintenance. Robots equipped with AI can learn specific preferences and adapt to the unique needs of a household. A Jetson-powered robot could be programmed to understand its owner's preferences, make decisions about how best to organize tasks, and even suggest changes to improve household efficiency based on patterns it detects over time.

Jetson Thor is also expected to play a significant role in personal assistant robots. These robots could interact with users using natural language processing, helping with daily tasks such as managing schedules, making phone calls, or controlling smart home devices. With its real-time processing ability, Jetson Thor enables robots to respond quickly and intelligently to user commands, making them feel more like human assistants than machines.

Additionally, in the area of elderly care, personal robots powered by Jetson Thor could provide critical support for seniors who wish to live independently. These robots could assist with mobility, monitor vital signs, and even provide companionship. AI capabilities would allow the robot to recognize potential emergencies, like falls or medical issues, and alert family members or medical professionals immediately.

As personal robotics continues to evolve, the sophistication of AI, paired with Jetson Thor's compact power, will make robots more versatile and capable of seamlessly integrating into people's daily routines. Whether in the form of smart home assistants or companion robots for the elderly, Jetson Thor is paving the way for more intelligent, adaptive robots that will make a significant impact on everyday life.

These applications of Jetson Thor demonstrate its potential to drive innovation across multiple sectors, from improving manufacturing efficiency to transforming healthcare and enhancing personal lives with smarter robots. With its combination of AI and advanced computing power, Jetson Thor is set to be at the heart of the next wave of robotics.

Chapter 6: Nvidia's OEM Strategy

Nvidia's strategy in the tech world has long been characterized by its focus on providing essential tools that empower developers, manufacturers, and enterprises to innovate. With its entry into the robotics market, the company has extended this approach by adopting an original equipment manufacturer (OEM) strategy, rather than competing directly with robot makers in the marketplace. This strategic shift is a natural extension of Nvidia's core business philosophy, which centers on providing high-performance, specialized components that enable other companies to build powerful systems.

OEM refers to a business model in which Nvidia supplies key components—such as chips, software, and platform technologies—that allow third-party manufacturers to build products that carry their own branding. In the case of robotics, Nvidia does not seek to become a robot manufacturer itself, but rather to provide the critical hardware and software that makes robots smarter, more efficient, and more capable. This includes Nvidia's famous Graphics Processing Units (GPUs), powerful AI models, and specialized hardware, such as the Jetson series, all of which are designed to be integrated into the robots produced by other companies.

One of the key advantages of Nvidia's OEM strategy is its scalability. By positioning itself as a supplier of components and technologies, Nvidia allows companies to leverage its powerful AI capabilities without requiring them to develop their own in-house systems from the ground up. This makes it possible for companies to focus on their core competencies—whether that be hardware design, user interface development, or robotic automation—while leaving the critical AI and processing power to Nvidia's advanced solutions.

Furthermore, Nvidia's ability to provide both hardware and software ecosystems makes its OEM strategy even more compelling. With platforms such as the Jetson series, Nvidia offers not just raw computing power but also the deep learning models, libraries, and simulation tools necessary to train robots in real-world environments. This results in a comprehensive ecosystem that significantly reduces the time-to-market for robot manufacturers.

Nvidia's OEM approach is also an effective way to tackle the growing demand for robotics across various industries. Rather than investing in the massive infrastructure required to develop robots in-house, Nvidia can focus on creating the most advanced AI and computational platforms, leaving the physical robots to be developed by a wide array of companies across multiple sectors, from manufacturing to healthcare. This allows Nvidia to target a much broader market than it would if it had to compete directly with every robot maker in existence.

Supplying the Tools, Not Building the Machines

Nvidia's decision to supply the tools rather than build the machines is a defining feature of its robotics strategy. In the world of technology, this approach is known as the "enabler model," where a company provides foundational infrastructure, platforms, and technologies that others can use to create more complex and market-specific solutions. By choosing this route, Nvidia is betting that its strength lies in the development of high-performance computing technologies, which can be universally adopted by a wide array of robot manufacturers rather than creating robots themselves.

Building the machines—especially humanoid robots—requires an immense investment in design, prototyping, and production that is outside of Nvidia's traditional wheelhouse. The company, by focusing on supplying the necessary computational power and AI models, can ensure its technology powers a broad range of robots without taking on the considerable costs and risks associated with physical product development. Nvidia's philosophy has always been to provide the best-in-class tools for solving complex computational problems, whether in gaming, data centers, or now, robotics.

The beauty of Nvidia's strategy lies in the versatility of its products. Nvidia's GPUs, with their parallel processing capabilities, are well-suited to the computational demands of robotics, which involve massive amounts of real-time data processing. The Jetson platform is another prime example, offering a specialized development kit and system-on-chip (SoC) solutions tailored to robotics. These tools are designed to power everything from autonomous vehicles to medical robots, providing the underlying infrastructure necessary for robots to "think" and learn in real-time.

By offering these powerful solutions as part of an OEM strategy, Nvidia enables third-party manufacturers to create custom robots without having to replicate the underlying hardware and software from scratch. For example, Tesla's humanoid robot, Optimus, uses Nvidia's technology as a foundational layer for its robotic systems. This allows Tesla to focus on the design and physical capabilities of the robot while relying on Nvidia to power its cognitive functions and AI capabilities.

This model also makes sense in terms of market growth. As robotics becomes more pervasive across industries, the number of companies developing robots is growing exponentially. Nvidia doesn't need to produce all the robots—it simply needs to ensure that its tools and technologies are the driving force behind these robots' intelligence. With the demand for robotics expected to soar in the coming years, Nvidia's tools will likely power a significant portion of the market, positioning the company as the go-to platform for developers and manufacturers.

Collaboration with Global Robot Makers

Nvidia's commitment to enabling robotics through collaboration with global robot makers is a crucial element of its success in the industry. Rather than attempting to disrupt the existing robotics ecosystem, Nvidia has made strategic moves to partner with companies across various sectors that are developing cutting-edge robots. By aligning with industry leaders, Nvidia gains access to a broader range of use cases for its technology, while manufacturers gain access to some of the most advanced AI and computing platforms available.

These collaborations are not limited to any one geographic region or market segment. Nvidia has established relationships with major players in industries ranging from industrial automation to healthcare and consumer robotics. For instance, companies like Amazon and Google, both of which are pursuing their own robot initiatives, use Nvidia's AI-powered platforms to drive the development of more autonomous and capable machines. These partnerships allow Nvidia to expand its influence in robotics without the need to directly compete with these companies in robot production.

The collaboration model works in a mutually beneficial way. Robot manufacturers gain access to Nvidia's powerful GPUs and AI models, which significantly enhance the capabilities of their robots, while Nvidia gains invaluable insights into how its technology is being applied across various use cases. This feedback loop drives continuous improvement, allowing Nvidia to refine and optimize its platforms, making them more versatile and adaptable to different industries.

These collaborations also extend to the development of specialized robotic solutions. For example, Nvidia has partnered with manufacturers of healthcare robots, helping them integrate advanced AI that can improve diagnostic capabilities, assist in surgeries, or provide care for elderly patients. In industrial settings, Nvidia's AI and GPUs are used in robotic arms for assembly lines, autonomous warehouses, and logistics solutions, enabling robots to work alongside human workers with increasing autonomy.

Nvidia's strategy of collaboration is deeply rooted in its understanding that robotics is a vast and fragmented industry. By supplying tools that meet the needs of a wide range of applications, Nvidia avoids the need to become a jack-of-all-trades robot manufacturer. Instead, it focuses on becoming the backbone of robot intelligence across the world's leading robotics companies.

The Market for "Hundreds of Thousands" of Robots

The concept of "hundreds of thousands" of robots is not merely a visionary statement, but a reflection of the growing reality in industries across the globe. Nvidia's OEM strategy is predicated on the understanding that the demand for robots—especially those powered by AI—will soon reach unprecedented levels. As automation and artificial intelligence continue to evolve, the potential for widespread adoption of robots is expanding rapidly across diverse sectors, ranging from manufacturing to healthcare and beyond.

The sheer scale of the market is driven by several key factors. First, robotics and AI are becoming increasingly accessible to businesses of all sizes. As the cost of developing advanced AI models and robotics systems continues to decrease, smaller companies and startups are entering the space, pushing the market size upward. Additionally, industries like logistics, healthcare, and manufacturing are increasingly turning to robots for tasks that were once done manually, creating a huge demand for robotic solutions.

Nvidia's tools—especially platforms like Jetson—are poised to capture a significant portion of this market. By offering the computing power needed for robots to perform complex tasks such as object recognition, motion control, and decision-making in real-time, Nvidia is setting the stage for the deployment of "hundreds of thousands" of robots globally. This number is not an exaggeration; it's a forecast of the future, based on the current pace of automation and AI adoption.

The rise of autonomous delivery drones, robots in agriculture, and personal assistants all point to the fact that robots are becoming integral to many aspects of modern life. Nvidia's collaboration with companies developing such robots ensures that its technology will be at the heart of this explosive growth. As robotics continues to scale, Nvidia's role as a provider of foundational AI and computing platforms will solidify, making the company a key player in the development of the next generation of intelligent machines.

Nvidia's focus on enabling a global ecosystem of robot makers, rather than competing with them, positions the company to capitalize on the rapidly expanding robotics market. The "hundreds of thousands" of robots predicted to enter the market are not just speculative—they represent the future of automation, with Nvidia's technology at the core of this transformation.

Chapter 7: Competitive Landscape

In the fast-evolving robotics and artificial intelligence (AI) sectors, the competitive landscape has been rapidly shifting, especially with advancements in AI-powered systems and robotics. Nvidia, a leader in the AI chip industry, has largely dominated the field with its high-performance GPUs (graphics processing units) and AI-centric technologies. However, as Nvidia intensifies its push into the humanoid robotics sector with Jetson Thor, it is clear that it faces significant competition from major players like Tesla, Amazon, and Google, each bringing its own strengths, investments, and innovations to the table.

Nvidia's primary strength lies in its ability to provide powerful hardware and software solutions to AI and robotics companies. Its Jetson platform, used in robotics, provides high-performance computing power that allows developers to create sophisticated, AI-driven robotic applications. However, the competition is increasingly heating up, with Tesla, Amazon, and Google each developing their own robotics platforms that rival Nvidia's offerings.

Tesla, for example, has been working on its humanoid robot, Optimus, for several years. Optimus, Tesla's AI-driven humanoid robot, is expected to serve both as a personal assistant and a labor force in Tesla's factories, helping with tasks like assembly line operations. Tesla's extensive experience in AI, coupled with its powerful in-house hardware and machine learning capabilities, makes it a formidable competitor in the robotics sector. Furthermore, Tesla's vertical integration model allows it to have greater control over both the hardware and software of its robots, reducing its reliance on third-party providers like Nvidia.

Similarly, Amazon has made significant strides in robotics, primarily focusing on automation in its warehouses. The company's use of AI-powered robots, such as the Kiva robots in its fulfillment centers, has significantly improved operational efficiency. With Amazon's vast resources and continuous investment in machine learning, AI, and robotics, the company has the potential to develop new robotic systems that could rival or even surpass Nvidia's offerings in specific niches, particularly in logistics and supply chain automation.

Google, through its subsidiary, Boston Dynamics, is another key player in the robotics industry. Boston Dynamics has gained significant attention for its agile, dynamic robots like Atlas and Spot. With a strong emphasis on mobility, AI, and machine learning, Boston Dynamics is focused on creating robots that can navigate real-world environments in ways that most robotic systems cannot. While its robots are not yet as commercialized as Tesla's or Amazon's, Boston Dynamics is at the forefront of developing advanced robotics technologies, making it a key competitor to Nvidia in the long run.

Nvidia's position in the robotics space is far from guaranteed. While the company offers the infrastructure and tools to help drive the industry forward, Tesla, Amazon, and Google each have unique capabilities that pose serious competition. Tesla's focus on building integrated hardware-software solutions for robotics, Amazon's dominance in logistics automation, and Google's expertise in AI and dynamic robot design create a multifaceted competitive landscape. For Nvidia to maintain its leadership, it will need to adapt quickly and respond to these innovations, ensuring that its offerings remain valuable and competitive.

Tesla, Amazon, and Google – The Major Players

Tesla, Amazon, and Google are undoubtedly the three biggest players in the robotics and AI spaces, each taking different approaches to leverage robotics in enhancing their business models and operations. While Tesla is focused on developing humanoid robots for industrial and consumer applications, Amazon uses robotics to optimize its massive logistics network, and Google, through its subsidiary Boston Dynamics, is revolutionizing robot mobility and AI.

Tesla

Tesla's entry into the robotics space is centered around its development of the humanoid robot, Optimus. Tesla's CEO, Elon Musk, has been vocal about the company's aspirations to create a robot that can perform tasks across various sectors, ranging from manufacturing and logistics to healthcare and personal assistance. Tesla's advanced work in AI, particularly its work on self-driving cars, gives it a unique advantage in the robotics sector. The company is well-positioned to integrate the power of AI with the hardware needed to build highly functional robots.

Tesla's approach is unique because of its deep integration of software and hardware. By building both the AI system (through Tesla's Autopilot technology) and the physical hardware (the robot's body, actuators, and sensors), Tesla has the ability to optimize and fine-tune its robots in a way that many of its competitors cannot.

Tesla's focus on creating a robot for the general labor market, rather than niche applications, sets it apart from many companies in the robotics space. Optimus is designed not only to be a helper in the factory but to be a versatile robot capable of performing tasks in everyday environments.

Amazon

Amazon's robotics efforts are heavily focused on enhancing its fulfillment and delivery operations. The company has developed a range of robotic systems, including the Kiva robots used in its warehouses, which autonomously move goods around the facilities to help reduce the time it takes to pick and pack items for shipment. The company also uses AI-driven systems to optimize its inventory management and delivery networks, ensuring that products are delivered as quickly and efficiently as possible.

Amazon's approach to robotics has been highly strategic. Instead of focusing on humanoid robots or general-purpose machines, the company's robotics efforts are directed toward highly specialized, purpose-built robots that solve very specific logistical challenges. This focus on automation in its warehouses has allowed Amazon to scale its operations faster than many competitors, using robotics to manage an enormous inventory across its global network. Additionally, Amazon's vast cloud computing infrastructure, including AWS, is integral in powering the AI systems that control these robots, making it another competitive edge the company has in this space.

Google

Google's robotics division, primarily through its acquisition of Boston Dynamics, is focused on creating robots with advanced mobility and autonomy. Robots like Spot, a quadruped robot, and Atlas, a bipedal humanoid robot, are designed to navigate complex environments with agility and ease. Google's focus is on creating robots that can perform tasks in environments where traditional robots might struggle, such as uneven terrain or navigating stairs.

Unlike Tesla and Amazon, Google has not yet integrated its robots into any specific commercial applications. However, its focus on mobility and AI is pushing the boundaries of what robots can do in the real world. Google has partnered with other companies and research institutions to further develop these robots, and it's likely that the company will continue to push the envelope in terms of robotics innovation, focusing on applications in logistics, military, and fieldwork in challenging environments.

How Competitors Are Responding to Nvidia

As Nvidia intensifies its push into robotics, its competitors have been quick to respond. Tesla, Amazon, and Google are all leveraging their existing technological infrastructures and R&D efforts to develop robotics solutions that will challenge Nvidia's dominance in the AI-powered robotics space.

Tesla's Response

Tesla's response to Nvidia's robotics push is centered on its own AI and hardware integration. Tesla's robotics ambitions, as part of its broader AI strategy, mean that it is increasingly less reliant on third-party providers for its hardware and software solutions. By developing the Optimus humanoid robot in-house, Tesla is creating a vertically integrated solution that competes directly with Nvidia's Jetson platform.

Tesla's approach is also significant in that it leverages its expertise in deep learning and neural networks developed for self-driving cars. This technology is particularly suited to the task of creating humanoid robots, which require sophisticated computer vision and decision-making systems. As a result, Tesla's response to Nvidia is not just about competing with hardware but also with AI software, a key strength of Nvidia's offerings.

Amazon's Response

Amazon has long been a leader in automation, but it is now making a more deliberate move toward robotics to improve its supply chain even further. Amazon's robotics arm is primarily focused on improving warehouse efficiency, and the company continues to develop AI and robotic solutions that reduce human labor in these environments. As Amazon seeks to scale its robotics capabilities, it is likely that the company will explore new avenues for humanoid robotics in customer-facing applications, making it a competitor to Nvidia in broader AI-powered robotics applications.

Google's Response

Google, through its acquisition of Boston Dynamics, is focusing on developing robots that can excel in real-world, dynamic environments. While it has not yet developed humanoid robots for mass production, the company is setting itself apart by creating robots with advanced mobility systems that can adapt to various environments. Google's response to Nvidia is thus more focused on the robotics hardware side, with an emphasis on agility, autonomous learning, and mobility. The development of Atlas and Spot illustrates the company's commitment to advancing robotics, making it a competitor to Nvidia in specialized robotic applications.

Will Nvidia Remain the Dominant Force?

The answer to this question depends on multiple factors: Nvidia's ability to adapt to new developments in the robotics sector, the continued evolution of AI technologies, and the competitive pressures from Tesla, Amazon, and Google. While Nvidia has a strong foothold in the AI and robotics industries, the company faces growing competition from companies with greater resources and unique strategic advantages.

Tesla, Amazon, and Google each have their own reasons for entering the robotics field, but they also bring their own strengths to the table. Tesla's vertically integrated hardware-software approach could give it an edge in the development of robotics that require tight control over both elements. Amazon's focus on logistics and warehouse automation may eventually push the company toward humanoid robotics, while Google's robotics arm is already testing boundaries with mobility-focused machines.

In the short term, Nvidia's dominant position in the AI chip market, particularly for machine learning and robotics applications, is likely to keep it at the forefront of the industry. However, long-term success will depend on Nvidia's ability to innovate, expand its ecosystem, and form strategic partnerships that allow it to stay ahead of its competitors.

Ultimately, the future will likely see greater collaboration between Nvidia and companies like Tesla, Amazon, and Google, rather than a cutthroat race to dominance. Nvidia's open-platform approach allows it to remain a critical player in powering the next generation of robotics, even as competitors develop their own hardware and software solutions. In this evolving landscape, Nvidia's ability to adapt to new market needs and maintain its leadership in AI technology will determine whether it remains the dominant force in robotics.

Chapter 8: The Future of Robotics and AI

The convergence of robotics and artificial intelligence (AI) marks one of the most transformative technological shifts of the 21st century. In the past few decades, we've seen rapid advancements in both fields, but this is only the beginning. Robotics has already revolutionized industries from manufacturing to healthcare, while AI is at the heart of this change, enabling machines to learn, adapt, and interact in ways that were previously unimaginable. As we look ahead, the future of robotics and AI is poised to shape everything from our economy and job markets to the way we live and interact with the world around us.

One of the most significant drivers of this future is the continuous advancement of machine learning (ML) and deep learning (DL) techniques. These technologies allow robots and AI systems to not just follow pre-programmed instructions but also to improve and refine their capabilities over time. Today's robots, such as those used in assembly lines or logistics, are already capable of performing tasks with impressive precision. However, the future will see robots that can make autonomous decisions based on real-time data and complex algorithms. These robots will be able to work alongside humans in a far more collaborative way, handling dangerous, repetitive, or physically demanding tasks, and allowing humans to focus on more creative and cognitive aspects of work.

AI will also play a critical role in the continued evolution of robotics. For instance, AI-enabled robots will be able to navigate and adapt to new environments with greater ease. The development of AI-powered vision and speech recognition systems will allow robots to interact with humans in increasingly natural ways. Think of home robots that not only clean but also understand your preferences, anticipate your needs, and respond to voice commands or gestures as seamlessly as a human assistant would.

Beyond the physical capabilities of robots, the integration of AI will allow them to perform complex decision-making processes. Autonomous vehicles are a prime example of this potential. In the future, AI will enable robots to process massive amounts of data in real-time, make quick decisions in high-pressure environments, and adapt to changing conditions without human intervention. While these technologies are already being implemented in pilot programs for autonomous vehicles, their widespread use is still in the early stages. In the coming years, it's expected that the infrastructure supporting AI-driven technologies will expand, making autonomous systems more reliable and prevalent in day-to-day life.

However, the road to achieving fully autonomous AI-powered robots will not be without challenges. Ethical concerns, particularly regarding job displacement and security, are at the forefront of discussions about the future of robotics. The potential for AI to replace human workers in certain industries is a real concern, leading to a debate about how society can balance technological progress with maintaining meaningful employment. Additionally, security risks involving autonomous robots are another critical issue. As robots become more connected and integrated into society, they will be more susceptible to hacking or manipulation by malicious actors. Ensuring the safety, accountability, and ethical use of AI and robotics will require rigorous standards and oversight.

As we look towards the future of robotics and AI, it's clear that we are entering a new phase where these technologies will not only augment human capabilities but will also serve as indispensable tools in enhancing efficiency, productivity, and even creativity. From healthcare to logistics, education to entertainment, the widespread implementation of AI and robotics is set to redefine our world.

What Comes Next for Humanoid Robots?

Humanoid robots, machines designed to resemble and replicate human actions, have captured the public's imagination for decades. From the sci-fi robots of old films to today's more advanced versions, humanoid robots hold a unique position in both the robotics and AI fields. These robots, designed to perform tasks similar to those humans can do, are often viewed as the ultimate expression of robotics and AI integration. As we look toward the future, several factors will influence the development and deployment of humanoid robots.

In the near term, humanoid robots will continue to make inroads in specific industries. Already, companies like Boston Dynamics and Honda have showcased robots capable of performing basic tasks like opening doors, lifting objects, and even dancing. While these robots are impressive, they are still largely in the experimental stage. The next major step in humanoid robot development will likely focus on improving their dexterity, balance, and autonomy. Advances in artificial intelligence and machine learning will enable humanoid robots to perform more complex tasks that require fine motor skills, problem-solving, and adaptability.

The healthcare industry is likely to be one of the first to widely adopt humanoid robots. Robots that can perform surgical procedures with precision, assist with patient care, or even act as companions for the elderly are all within reach in the coming years. Humanoid robots could serve as assistants in the operating room, helping human surgeons perform delicate tasks or monitoring patients for signs of distress. In home care, humanoid robots could assist elderly or disabled individuals with daily activities, offering support that is personalized to the individual's needs. For instance, a humanoid robot could help an elderly person take medication on time or provide reminders for doctor's appointments.

Another key application of humanoid robots will be in customer service. We've already seen robots like SoftBank's Pepper being deployed in retail and hospitality environments. In the future, humanoid robots will likely become even more integrated into the service industry, offering personalized customer service, answering questions, and even processing transactions. As humanoid robots become more human-like in their interaction and communication capabilities, they may eventually become indistinguishable from human workers in certain service roles.

However, widespread deployment of humanoid robots will not be without its challenges. One major obstacle is cost. Building robots that closely mimic human form and function is expensive, and for many industries, the return on investment may not be high enough to justify the expense. Another challenge lies in the social acceptance of humanoid robots. While many people are comfortable with machines that perform specific tasks, the idea of robots that look and behave like humans raises ethical and emotional questions. How will people feel about robots caring for their children or elderly parents? Will we accept robots as members of society, or will they always be viewed as tools or curiosities?

Despite these challenges, the continued development of humanoid robots will likely be fueled by advancements in AI and robotics. As the cost of materials and manufacturing decreases and AI systems become more efficient, the potential for humanoid robots to play a significant role in society will only grow. In the long term, humanoid robots could become integral members of our workforce, contributing to fields like healthcare, education, entertainment, and service.

AI's Role in Shaping Everyday Life

Artificial Intelligence is no longer a futuristic concept; it's already a part of our everyday lives. From personal assistants like Siri and Alexa to recommendation systems on Netflix and Amazon, AI is helping us make decisions, organize our lives, and even entertain us. As AI continues to evolve, its influence on daily life will become even more pervasive and profound.

In the realm of personal convenience, AI is already making a significant impact. Smart home devices powered by AI are transforming how we manage our homes. Thermostats learn our schedules and adjust temperatures accordingly, while security cameras use AI to detect unusual activities. In the future, AI-powered home assistants will become more intuitive, understanding not only what we need but also predicting our desires before we express them. For instance, AI systems could anticipate when we'll need to make dinner and suggest recipes based on what's in our fridge.

In the workplace, AI is revolutionizing industries by automating routine tasks and providing valuable insights. In the next decade, AI will likely become even more integrated into the workplace. Machine learning algorithms will continue to assist in decision-making, while AI-powered tools will handle more complex tasks such as data analysis, project management, and customer service. This will enable employees to focus on higher-level creative tasks, while AI will handle the heavy lifting of data processing, pattern recognition, and routine tasks.

The education sector is another area where AI is having a profound impact. AI-powered tutoring systems can provide personalized learning experiences, adapting to each student's unique needs and progress. In the future, AI could even design custom curriculums for students, ensuring that each individual receives the most effective and efficient education based on their strengths and weaknesses. Teachers will be empowered with AI tools to track student progress, offer personalized feedback, and even identify at-risk students before they fall behind.

Healthcare is one of the most promising fields where AI will have a life-changing impact. Already, AI is being used for diagnostic purposes, identifying diseases from medical images with greater accuracy than human doctors. As AI continues to improve, it could revolutionize preventive care by analyzing genetic information, lifestyle choices, and environmental factors to predict health risks before they become serious conditions. In the future, AI could even assist in the development of personalized treatment plans, ensuring that every individual receives the care best suited to their needs.

While the role of AI in everyday life will bring immense benefits, it also raises important questions about privacy, ethics, and job displacement. The widespread use of AI-powered systems will require careful consideration of these issues to ensure that AI is used responsibly and fairly.

Predictions for the Next Decade

Looking ahead to the next decade, the pace of technological advancement is expected to accelerate exponentially. By 2034, we could see a world where AI and robotics are seamlessly integrated into every aspect of our lives. Here are some key predictions for the next decade:

1. **Autonomous Vehicles:** We will likely see widespread adoption of autonomous vehicles, transforming the transportation industry. By 2034, autonomous cars, trucks, and drones could

become the norm, reducing accidents, improving traffic flow, and making transportation more efficient.

2. **AI and Healthcare:** AI will play a major role in revolutionizing healthcare by providing faster, more accurate diagnoses, personalized treatments, and even predictive health monitoring. We could also see AI-driven drug development, potentially cutting down the time it takes to bring new medications to market.

3. **The Rise of Virtual and Augmented Reality:** VR and AR technologies will mature, enabling new forms of entertainment, education, and work. By 2034, VR could be as common as smartphones are today, changing the way we interact with digital content.

4. **Automation and Jobs:** Automation will continue to reshape the workforce. While some jobs will be replaced by machines, new types of

employment will emerge, particularly in fields like AI ethics, machine learning, and robot maintenance.

5. **AI in Creativity:** AI will no longer just assist in repetitive tasks; it will play a role in creative endeavors. We could see AI creating art, writing novels, and even composing music, challenging our traditional notions of creativity.

The next decade promises to be a transformative period in which AI and robotics will become deeply woven into the fabric of everyday life. The key will be balancing innovation with ethical considerations, ensuring that these technologies are used responsibly for the benefit of all.

Chapter 9: Challenges and Ethical Considerations

The rapid advancement of AI-powered robots, particularly in industries like manufacturing, healthcare, and service sectors, presents both immense opportunities and significant ethical challenges. As these technologies evolve, so too does the need for a comprehensive framework to address the societal, economic, and moral questions they raise.

One of the foremost challenges with AI-powered robots is the risk of exacerbating existing inequalities. Automation and robotics can potentially concentrate power and wealth in the hands of those already controlling the technologies, thus further marginalizing individuals and regions that lack access to these innovations. This issue, often referred to as the "automation divide," suggests that while large corporations and wealthy nations will reap the benefits of AI, the disadvantaged may face even greater disparities in access to employment opportunities and resources.

Ethically, there's also the question of human agency. As robots become more capable of performing tasks autonomously, the role of human workers and decision-makers is increasingly being called into question. The shift towards machines taking over complex tasks requires society to consider not only the economic impact of automation but also its psychological and social effects. People may feel displaced or marginalized, raising concerns about the loss of human dignity in the workplace and the erosion of social value attached to work.

Another challenge lies in ensuring that AI systems reflect ethical decision-making frameworks. As robots begin to make choices independently of human operators, we must ask whether they are capable of acting in ways that align with societal norms and values. For instance, self-driving vehicles must make decisions during emergencies, such as choosing between the safety of a pedestrian or the passengers. How do we ensure that AI robots make decisions that society deems morally acceptable? And how do we design robots that can interpret complex human emotions or the nuances of a given situation? These ethical considerations are still being debated in academic and industry circles.

The Ethics of AI-Powered Robots

AI-powered robots raise fundamental questions about autonomy, responsibility, and the role of artificial intelligence in human society. One of the most pressing ethical issues is the concept of autonomy. As robots become increasingly sophisticated, the question arises: when does a robot have enough decision-making power to be considered autonomous, and how do we handle its accountability? For example, if a robot performs an action that results in harm, is it the manufacturer, the user, or the robot itself that should be held responsible?

Autonomy in AI-powered robots introduces a slippery slope in terms of moral accountability. If an autonomous robot makes an error—say, in a medical diagnosis or an industrial operation—who should be held accountable? This issue has been at the forefront of debates surrounding self-driving cars and healthcare robots. For instance, if a robotic surgery tool makes a mistake during an operation, is the robot to blame, or does responsibility fall to the developers, the medical professionals who used it, or both?

Another ethical concern lies in the idea of "robot rights" or the rights of autonomous machines. While this may seem like science fiction, discussions are beginning to surface about the future of robot personhood. Should highly advanced robots, particularly those capable of emotional intelligence, have rights similar to human rights, or are they simply tools created to serve human needs? These questions challenge existing legal systems, which are not equipped to consider robots as entities that could possess rights, despite their growing capabilities.

Furthermore, there is the ethical issue of the misuse of AI-powered robots. The potential for these robots to be weaponized or deployed in ways that infringe on privacy or human rights is a legitimate concern. For example, AI-powered surveillance robots can easily be used for mass surveillance or to suppress dissent, potentially violating civil liberties. In a military context, autonomous weapons systems could make decisions without human oversight, leading to unintended consequences.

Finally, there's the issue of bias in AI. Since AI systems learn from data, they are often prone to inheriting biases present in the data they are trained on. If the data used to train AI robots is biased, the robots may act in ways that perpetuate these biases. This can have devastating consequences, particularly in sensitive applications like criminal justice or hiring. For instance, a biased robot deployed to evaluate candidates for a job might unfairly favor one demographic group over another, perpetuating existing societal inequities.

Addressing Job Displacement and Automation

Perhaps the most discussed ethical consideration surrounding the rise of AI-powered robots is job displacement. As robots and AI systems increasingly take over tasks traditionally performed by humans, fears of widespread unemployment and economic dislocation have grown. While many industries have already seen the integration of robots for tasks like assembly line work, this trend is rapidly expanding into more complex sectors, such as healthcare, transportation, and even education.

The displacement of workers by automation raises significant questions about the future of work. Jobs in sectors like manufacturing, retail, and logistics are at the highest risk of automation, but many white-collar jobs may also be affected as AI-driven solutions become more capable of complex tasks, including legal research, customer service, and data analysis. What happens to workers who no longer have the skills required for these new roles? And how does society support these individuals as they transition into a new workforce?

In many ways, addressing job displacement through automation requires a multifaceted approach. One solution is the concept of retraining and reskilling. As robots take over routine tasks, there will be a need for new types of jobs—roles that require uniquely human skills like creativity, empathy, and complex problem-solving. However, retraining and reskilling programs must be widely accessible and tailored to those most at risk of losing their jobs to automation. Without these initiatives, large segments of the population could find themselves unemployable in the new economy.

Another approach is the consideration of universal basic income (UBI), a concept that has been gaining traction as a solution to automation-induced job displacement. UBI proposes providing all individuals with a guaranteed income, regardless of employment status, as a means to ensure basic financial security in an increasingly automated world. However, UBI remains a controversial solution, with critics arguing that it could disincentivize work and create a dependency on government support.

At the policy level, governments must find ways to balance the benefits of automation with the need to ensure that economic growth is shared equitably across society. This involves creating policies that not only support innovation and technological advancement but also protect workers and ensure fair economic outcomes.

Ensuring Safety and Accountability

With the growing reliance on AI-powered robots, safety and accountability have become paramount concerns. Robots, particularly those operating autonomously in environments such as hospitals, factories, and public spaces, pose risks if they malfunction or behave unpredictably. As these robots become more integrated into critical sectors, ensuring their reliability and safety becomes crucial not only for the success of the technology but also for the well-being of human users and bystanders.

First and foremost, there is the challenge of ensuring that robots are programmed to follow clear and strict safety protocols. In industrial environments, for example, robots must be able to detect and avoid human workers to prevent accidents. The failure to implement adequate safety measures could result in serious injuries or even fatalities. This extends to self-driving vehicles, where a malfunction could lead to a collision or loss of life.

The accountability issue is equally pressing. If a robot malfunctions and causes harm, who is responsible? As mentioned earlier, the question of accountability is murky when it comes to autonomous robots. Who is legally responsible for a robot's actions: the developer, the manufacturer, or the end-user? These questions have significant implications for liability, insurance, and legal systems. For instance, in the case of a self-driving car accident, determining fault becomes increasingly complex, especially when the car's AI made a split-second decision that could be interpreted in multiple ways.

Furthermore, there is the issue of monitoring and oversight. Even the most advanced robots should be subject to rigorous oversight to ensure they are operating within ethical and safety guidelines. Regulatory bodies and government agencies must establish clear frameworks for evaluating and certifying AI-powered robots. These frameworks must evolve alongside technological advancements, ensuring that safety standards are consistently updated to account for new developments in AI and robotics.

Finally, as robots become more integrated into daily life, the need for public confidence in their safety becomes crucial. Transparent communication from manufacturers about the potential risks and safety protocols will help build trust. This, in turn, will encourage broader adoption of AI-driven technologies while mitigating fears of accidents or misuse.

Conclusion

Jetson Thor represents more than just an evolution of Nvidia's robotics strategy; it is a pivotal moment in the ongoing robotics revolution. As the world continues to embrace automation and AI, the role of compact, powerful computing systems like Jetson Thor will become central to the development of humanoid robots. Unlike traditional approaches, Nvidia is not attempting to manufacture the robots themselves, but rather offering the tools that enable countless robot makers to build and innovate. This shift from product manufacturing to supplying high-performance hardware for the masses will democratize the field of robotics, opening up opportunities for small and medium-sized companies to enter the market.

With its cutting-edge AI processing capabilities, Jetson Thor provides an essential platform for building robots that can learn, adapt, and perform complex tasks. The system is designed to handle generative AI models, which are crucial for training robots in simulated environments. This advancement is particularly significant for humanoid robots, which rely on complex decision-making processes and sophisticated sensory input to operate effectively in real-world environments. By equipping robot manufacturers with Jetson Thor, Nvidia ensures that the next generation of robots can be faster, smarter, and more adaptable, propelling the industry forward.

Furthermore, Nvidia's vision of supplying the "hundreds of thousands" of robot builders around the world aligns with the growing demand for robots across various sectors. Whether in manufacturing, healthcare, or personal service, the integration of AI-powered robotics will continue to revolutionize industries. Jetson Thor's place in this transformation is crucial—it represents both the present and future of robotics, a testament to Nvidia's strategy of empowering others to innovate while simultaneously shaping the future of AI-driven automation.

The future of AI and automation is one marked by rapid technological advancements and societal transformations. As we look ahead, one thing is clear: AI will continue to play a dominant role in reshaping industries and our daily lives. The marriage of AI and automation, powered by platforms like Nvidia's Jetson Thor, will drive innovation across fields such as manufacturing, healthcare, logistics, and even entertainment. But as AI systems become more integrated into our environment, several key factors will dictate the pace and nature of this revolution.

First, AI's capabilities will continue to expand, with deep learning models becoming more sophisticated and capable of handling complex tasks with greater efficiency and accuracy. The use of generative AI, in particular, will allow robots to learn from simulated environments, vastly improving their ability to navigate real-world scenarios. With these advancements, we are poised to see more autonomous systems that can operate independently, making decisions and solving problems without direct human input.

Second, automation will continue to reduce human involvement in repetitive, dangerous, or low-skill tasks. This will lead to significant productivity gains and cost reductions in sectors like manufacturing and supply chain management. However, the rise of automation will also raise critical questions about the future of work. While automation will create new opportunities in areas like AI development, robotics maintenance, and system design, it will also necessitate a focus on reskilling workers who may be displaced by these technologies.

Lastly, the ethical and societal implications of AI and automation will need to be carefully managed. Issues related to privacy, security, and fairness will become more pronounced as AI systems handle sensitive data and make decisions that affect people's lives. The challenge will be to ensure that these technologies are deployed responsibly, with frameworks in place to protect individuals and society as a whole.

Ultimately, the future of AI and automation will be defined by balance—advancing technology while safeguarding the well-being of humanity.

www.ingramcontent.com/pod-product-compliance
Lightning Source LLC
Chambersburg PA
CBHW062109220526
45471CB00010B/3658